OFF GRID SOLAR POWER SIMPLIFIED

THE DIY GUIDE TO INSTALL A MOBILE SOLAR POWER SYSTEM IN BOATS, RVS, VANS AND TINY HOMES

By

WILL SMART

© **Copyright 2020 By Will Smart**
All rights reserved.

This document is geared towards providing exact and reliable information with regards to the topic and issue covered. The publication is sold with the idea that the publisher is not required to render accounting, officially permitted, or otherwise, qualified services. If advice is necessary, legal or professional, a practiced individual in the profession should be ordered.

- From a Declaration of Principles which was accepted and approved equally by a Committee of the American Bar Association and a Committee of Publishers and Associations.

In no way is it legal to reproduce, duplicate, or transmit any part of this document in either electronic means or in printed format. Recording of this publication is strictly prohibited and any storage of this document is not allowed unless with written permission from the publisher. All rights reserved.

The information provided herein is stated to be truthful and consistent, in that any liability, in terms of inattention or otherwise, by any usage or abuse of any policies, processes, or directions contained within is the solitary and utter responsibility of the recipient reader. Under no circumstances will any legal responsibility or blame be held against the publisher for any reparation, damages, or monetary loss due to the information herein, either directly or indirectly.

Respective authors own all copyrights not held by the publisher.

The information herein is offered for informational purposes solely, and is universal as so. The presentation of the information is without contract or any type of guarantee assurance.

The trademarks that are used are without any consent, and the publication of the trademark is without permission or backing by the trademark owner. All trademarks and brands within this book are for clarifying purposes only and are the owned by the owners themselves, not affiliated with this document.

TABLE OF CONTENTS

INTRODUCTION -------- 5

FIND OUT HOW SOLAR ENERGY WORKS -------- 13

SELECTION OF A SOLAR ENERGY SYSTEM -------- 18

WHAT IS A SOLAR SYSTEM? -------- 21

DIY SOLAR PANELS FOR HOME USE: ELIMINATES HIGHER BILLS -------- 23

DIY SOLAR POWER SYSTEM COMPONENTS -------- 26

INSTALLATION OF A SOLAR PANEL ON-BOARD YOUR BOAT -------- 30

SOLAR PANELS FOR BOATS -------- 39

DIY SOLAR POWER GENERATION TIPS FOR YOUR BOAT -------- 41

SOLAR ENERGY FOR YOUR MOTORHOME -------- 44

ALTERNATIVE POWER SUPPLY FOR CAMPERS -------- 47

THINGS TO KNOW ABOUT RV SOLAR POWER ----------- 51

DIY SOLAR MOTORHOME SYSTEMS - HOW CAN I CREATE A DIY SOLAR MOTORHOME KIT? ---------------- 54

THE FOUR ESSENTIAL COMPONENTS OF A SOLAR MOTORHOME SYSTEM AND WHAT THEY DO ----------- 57

HOW SOLAR ENERGY WORKS: ON-GRID, OFF-GRID AND HYBRID SYSTEMS --- 68

SIMPLE INSTRUCTIONS FOR INSTALLING SOLAR ENERGY IN THE VAN -- 78

HOW TO INSTALL SOLAR PANELS FOR STEP-BY-STEP INSTRUCTIONS FOR YOUR SMALL HOME ---------------- 86

HOME SOLAR SYSTEM - DIY SOLAR PANELS ---------- 101

TEN REASONS TO INSTALL A SOLAR-POWERED PUMPING SYSTEM --- 108

INTRODUCTION

For a long time, solar energy has been identified only as the conversion of sunlight into electricity. Although this is not incorrect, most of us do not know that the energy obtained must first be converted into electricity to generate generally functioning electricity. The conversion is made possible by photovoltaics, a method in which semiconductors are used to convert solar radiation into electricity.

Segments of a straightforward Solar power system:

Solar or photovoltaic cells

The get together of semiconductors and hardware, or solar power cells is remembered for a photovoltaic module, usually known as a solar power board. Several solar modules are called solar module arrays.

Battery

A solar panel-based board gathers and produces vitality from solar radiation. Photovoltaic modules convert this energy into direct current, a current generated by cells. Although it is possible to connect a direct current load directly to the solar panel, batteries play an essential role in a properly functioning photovoltaic system.

Regulator

A controller is optional and still constitutes an essential component of a photovoltaic system. During the cyclical process, the battery is likely to be overcharged or discharged, reducing battery life. A regulator prevents such situations by regulating the condition of the batteries. It maintains a charge level that monitors when the battery is overcharged or discharged. Generally, a regulator keeps the batteries in the most suitable working conditions.

Converter / Inverter

A converter converts the direct current stored in the batteries into alternating or irregular current, the type of energy used by the mains power supply. DC/AC or direct/inverter is also known as an inverter and is used to adapt the current and voltage required to the load. It is typical for energy loss to occur during power conversion.

Load

All devices that consume electricity are considered charged.

While picking a heap for the solar power system, it is vital to begin a low vitality segment before introducing extra sunlight-based modules to abstain from squandering resources. Photovoltaic systems are ideal for lighting because lights consume only a few watts compared to larger devices such as televisions, components, or computers. Some lights operate on direct current and allow the use of solar energy systems in a small part.

Solar collectors, batteries, regulators, converters, and loads form the solar collector system. If all these components are installed correctly, a solar poper system can support itself for years.

Advantages of Solar Energy

For most home users, the electricity grid is the most convenient source of electricity. It appears that a serious blackout or power outage is still in the distant future. However, growing concern about fossil fuel depletion is prompting authorities around the world to use renewable energy sources. Since solar energy is the most commercialized among other renewable energies, it is necessary to know how you, as end-users in residential areas, can benefit from solar energy.

Economic Production

Solar energy is a proven commercial energy source. In addition to other renewable resources such as wind, water, biomass, biofuels, and geothermal energy, solar energy is the only clean energy capable of generating a large market, including private users. Due to advances in solar technology and the resulting improvement in financial approaches, the implementation of solar energy projects is steadily falling.

China's rise to become one of the largest solar module manufacturers has a significant impact on the cost of consumables. The nation additionally creates wind turbines, although fares are restricted, rather than world creation and the export of photovoltaic modules.

Infinite Energy Source

The extraction of fossil fuels is a constant threat to energy security. Fossil fuels are non-renewable sources of energy, and their depletion is inevitable, leading to supply shortages worldwide. On the other hand, renewable energies such as the sun, wind, and geothermal vitality can produce power without draining natural assets. These natural energies are limitless sources of vitality. With correct and strategic installation and use, renewable energies can supply clean electricity to the whole world.

Earth-Friendly

Like any green energy, solar energy produces relatively small quantities of greenhouse gases; one of the main factors in the depletion of the ozone layer. The use of alternative energies also promotes the production of low carbon technology products such as LED lights, low carbon devices, and hybrid cars. More energy-efficient storage tanks and solar collectors have been developed to improve the vitality of alternative energies.

Change in Electricity Consumption

Solar power plants are not only able to generate and supply electricity for private users, but also allow end-users to modify their electricity consumption. This is made conceivable by a two-way savvy, lattice, framework between the principle power provider and the shopper. If your solar power system generates excess energy, the power meter turns backward. An asynchronous inverter is required because it adapts to the main incoming power. If such favorable conditions occur, the electricity supplier will refund you the excess energy generated by the solar power system. Alternative energy is inherently unpredictable because the amount of energy produced depends heavily on weather conditions.

Relentless Government Support

The fact that the recession has had a limited impact on the demand for alternative sources shows that renewable energy is a stable and continuously growing industry. The tireless government support, including incentive packages from several countries, has boosted the sector, particularly in the production of solar, wind, and biofuels. Smart energy technologies are increasingly supported by capital and private equity investors and are giving way to digital and energy-saving applications on the market.

Governments around the world offer tax credit incentive packages and incentives for private, commercial, and industrial users. In addition to tax deductions for individuals and companies that install solar energy systems, the federal government also offers cashback bonus programs, exemptions from property tax, exemptions from sales tax and incentives for electricity companies. Investors also work with solar companies to promote recycling programs and allow conscious consumers to properly dispose of old products.

Low Maintenance and Operating Costs

The ideal setting requires optimal sun exposure during the day. If this is achieved, the constant generation of energy can be expected in perfect weather conditions. However, adequate operations and maintenance must be performed regularly to ensure optimal sunlight collection.

With correct and strategic installation, the photovoltaic modules are practically maintenance-free. Essential maintenance of solar modules includes cleaning the solar module and dirt. We also recommend washing photovoltaic modules, especially if you live in a particularly dusty region. Use non-abrasive cleaners and wash the cloth to avoid scratching the plates.

Life Expectation

Aside from their self-sufficiency, solar modules have an average lifespan of 20 or more years. With the latest developments in materials for the construction of photovoltaic modules, the life expectancy and profitability of solar modules are expected to improve in the coming years. In these years, approximately $2,000 is needed as maintenance and operating costs for photovoltaic modules.

Eliminate the Costs and Difficulties of Conventional Fuel Transportation

The federal government is allocating billions of dollars to transport fuel and other natural gas to generate electricity in the country. Solar energy systems reduce these costs because photovoltaic modules do not need fuel or natural gas to convert sunlight into electricity. The construction of large solar power plants also allows the local generation of ecological electricity, which can be supplied to private, commercial, and industrial users.

Solar Lighting

Internal solar lighting is possible through a system that collects and distributes sunlight for internal lighting. External solar lighting consists of a simple solar energy system in which the lights are continuously charged during the day and discharged at night by illuminating pedestrian crossings, for example. Lighting consumes a lot of energy and is therefore quite expensive in comparison to solar lights, which have become an efficient and convenient green option. Solar lights have a long-range and include outdoor solar floodlights, pier lights, Brinkmannsolar lights, and even solar-powered night lights. Solar lights are all the rage today because hundreds of different products and types are marketed. Solar lighting is now used inside the house and also outside, including the roof and garden! Even the commercial space today is equipped with a solar lighting system that not only pays in terms of cost savings but also helps to reduce the carbon footprint, which is significantly improved by current electricity consumption.

FIND OUT HOW SOLAR ENERGY WORKS

The fireball glowing in the sky, which we call the sun, is an infinite supplier of abundant radiant energy. This radiation energy, which is also called solar energy, is made up of electromagnetic waves. Humans have always been looking for new forms of energy. As a result, humans have been able to invent some very effective means of capturing part of this radiant energy with sophisticated means and converting it into various other useful forms of energy such as heat and electricity. This usable energy from the sun creates the solar energy we hear about today. Today solar energy is widely considered the most promising alternative energy source for the future and the topic of solar energy arouses great interest. This section will cover many of the essential aspects of solar energy so that we can better understand its true nature and understand its importance in the modern world.

Find Out How Solar Energy Works

Investigate, and you will, without a doubt, see that sun powered vitality is used in a huge number of ways in your everyday environment. Solar energy is rightly advertised as the next big news in the power generation, along with wind, hydroelectric power, and other alternative energy sources. It is, therefore, essential to learn more about solar energy and how to avoid an impending energy crisis in the not too distant future. Solar energy is transmitted through its rays, which contain abundant electromagnetic waves. Did you know that about 70 percent of the total amount of radiation received from earth is absorbed by the earth's surface, water, and vegetation? The remaining 30 percent is usually reflected into space.

The radiation that is absorbed by the earth's surface is responsible for heating the atmosphere and generates the so-called radiant heat. Solar radiation is directly responsible for about 99% of the usable flow of renewable energy on earth. However, the actual extent of the use of solar energy, as well as the costs for solar energy, depend primarily on the efficiency with which the established radiation is used by existing solar energy technology.

Although solar energy has several applications, its actual use is determined by the specific requirements it is intended to meet and by the various techniques used to capture and convert solar energy. Depending on the use and application, solar energy can be roughly divided into two categories: Active solar energy and passive solar energy. Vibrant solar energy uses sunlight to charge photovoltaic cells and solar collectors, which in turn generate various other forms of energy (mainly heat, electricity, and mechanical energy). The way it works is to charge solar cells with radiant energy and generate the electricity used to operate the pumps and rotating fans, which in turn generate other forms of energy for human consumption. When passive solar energy is used, sunlight is used to control the design of buildings with well-lit and well-circulating rooms, with a building that optimally overlooks the sun, and so on.

Active or passive, regardless of the form in which it is used, the advantages of solar energy are many. Today there is no doubt that solar energy has immense potential to replace conventional electricity and prove to be an independent and 100% sustainable alternative. Indeed, solar energy has touched many different aspects of our lives that we might not even realise. Solar energy has found its use in a variety of sectors, such as B. Family, business, urban planning and architecture, agriculture (including horticulture and greenhouses), solar lighting, solar heating, ventilation and cooling, disinfection and desalination, kitchen, electricity and so on. So, the importance and acceptance of solar energy is increasing greatly in today's world. One must also bear in mind that solar energy also has some disadvantages. Let's take a look at them.

Here are some Advantages and Disadvantages of Solar Energy

To better understand solar energy and how it can help us, we need to understand the various advantages and disadvantages of using it. The main advantage of using solar energy, which far outweighs possible disadvantages, is above all, the clean and environmentally friendly nature. Contrary to conventional power plants, the generation of solar energy does not produce harmful by-products, nor does it release harmful or polluting gases into the atmosphere. Furthermore, it does not depend on the already diminishing reserves of our natural resources, such as coal and oil. Also, the sun is a constant source of energy that never runs out. No wonder solar energy makes its way into more and more areas of our lives every day. However, when it comes to disadvantages, you need to keep in mind that solar energy requires a constant supply of strong sunlight to be truly effective. Some areas may not be able to receive the necessary amount of sunlight. However, research continues in this area to overcome these obstacles and make solar energy a more practical choice for everyone.

Solar energy is the ideal choice for you if you are serious about going green and also want to save a lot of money on electricity bills. The best thing about solar technology is that it can be implemented with the simplest means. Building solar panels in the home requires a certain level of familiarity with the subject. The internet is a great place to find many resources to get started. However, your chances of success for your project improve if you have detailed instructions. This way, you can make sure that there are no obstacles and that you get the best results.

SELECTION OF A SOLAR ENERGY SYSTEM

Using solar energy, you can generate electricity to power your home, be it a house or a mobile home. Many systems are available for generating solar energy. It often happens that the right choice is obvious. Still, in other circumstances, the decision as a whole can be more difficult and requires that the advantages and disadvantages of each system be carefully weighed. This section will guide you through the options available and explain what factors should be considered.

First, let's consider a mobile home. When choosing a solar power system for the mobile home, portability and flexibility are the most important factors. In general, standalone portable systems are the best solar power systems for your mobile home. These systems are easy to transport and can be flexibly aligned in different directions to take advantage of positive weather conditions.

If you are thinking about solar energy for your home or

business, you should consider an autonomous system. The typical structure of this system is to connect one or more solar modules to a battery—the battery stores direct current from the solar panel. Most devices use AC or DC, so an inverter is used to convert direct current into alternating current.

For people who are not sure about solar energy or who live in a colder climate, there is a so-called grid-connected solar energy system. This solution is a mixture of conventional electricity and solar energy. The property remains connected to the electricity grid but also has a solar panel. That is, if the sun does not shine, and you do not have enough electricity stored, you still have the power grid as a safety net.

Depending on the climate, many people start using grid-connected systems before finally switching to 100% solar energy, as it is significantly cheaper and is better for the planet than using the grid. As soon as a property is converted to 100% solar energy, a second solar power system is often installed. As a result, you can maintain the first system without interrupting the power supply and also have a system to act as a backup.

Currently, 100% of the networks are in the minority, except, of course, in areas where there is no network. If current trends continue, however, this scenario appears to appeal to home and business owners in hot climates who increasingly opt for 100% solar energy.

WHAT IS A SOLAR SYSTEM?

Over one million solar systems have been installed in Australia.

A solar power system consists of several photovoltaic modules (photovoltaic modules), a DC-AC converter (called an inverter), and a rack system that holds the photovoltaic modules in place.

Photovoltaic solar modules are generally mounted on the roof. They should be facing east, north, or west. The panels must be inclined at certain angles to maximize solar radiation on them.

Solar photovoltaic modules on the roofs of homes and businesses generate clean electricity by converting energy into the sunlight. This conversion takes place inside solar panels from specially manufactured materials from which solar panels are made. It is a process that requires no moving parts. In most cases, the solar modules are connected to the electricity grid using a device called a solar inverter.

Solar panels are different from solar hot water systems, which are also mounted on domestic roofs but use the heat from the sun to provide domestic hot water. In principle, similar to a hose in summer, it contains hot water in the sun after a few hours.

The innovation to change over sunlight into power was created in the nineteenth century; however, it was uniquely in the second half of the twentieth century that the advancement behind the need to give dependable remote force supplies and space satellites for families and businesses in Australia has accelerated.

Solar boards have been introduced on the tops of houses and other buildings in Australia since the 1970s. In 2013 the Clean Energy Council celebrated the fact that over 1,000,000 solar panels have been installed across Australia.

DIY SOLAR PANELS FOR HOME USE: ELIMINATES HIGHER BILLS

Information is spreading rapidly on the internet, and more and more people are becoming familiar with do-it-yourself solar modules. Given the changing world climate and unpredictable weather, many people are giving solar energy a second thought.

As weather conditions have changed, heatwaves and unusually cold winters are common. The unusual climate leads to an increase in energy consumption, which often leads to power outages. The sun is again becoming a viable alternative to rising energy costs and unreliable electricity grids. In the past, the only way to get a solar system was by the professional installation. The average person cannot afford the expensive costs associated with this system. With the ability to build your system, however, it is achievable for most people.

Building DIY solar panels for home use can only cost $200. The most expensive part of the system is the solar cell, which you can find on many popular auction sites at a fraction of the cost.

These cells can be easily damaged, but they are still very functional. Another option is to find a motorhome dealer near you. They sell camper accessories including solar panels. Usually, you can buy a panel there for around $100.

An instruction manual containing details of the system structure is required. They usually cost between $30 and $60. You can also find cheaper versions of eBooks, which often include video instructions. These instructions are very detailed, easy to follow, and provide a complete list of all the materials needed to build the system. Once you have the solar panel, you can find the rest of the materials in any hardware store.

The system should take no more than two days to be ready for use. This depends on the number of solar modules required. The instruction manual contains the necessary information to help you determine the number of panels required for your home. Most houses need at least two panels and usually no more than three.

The transition to solar energy is easier than ever. Do-it-yourself solar panels for the home are the cheapest way to install a solar system that can save money by protecting the environment. You can compensate for your investment in a few months, and you are free from the increase in electricity bills.

DIY SOLAR POWER SYSTEM COMPONENTS

With today's energy costs, solar energy is becoming increasingly popular. The industry grows by over 30% per year. While it costs a lot to control a whole house with a monetarily installed system, much can be saved by starting from the beginning and doing it yourself. Even the smallest system helps reduce the electricity bill.

Solar energy systems can be classified as connected to the grid or independent of the grid. Grid networks are connected only to the local power line and provide electricity when solar modules produce electricity. Off-grid systems are not connected to an electricity supplier and have a battery bank for storing energy so that energy can be supplied when the sun is unavailable. There are also hybrid systems that are connected to power lines and have a battery compartment.

The various components of a solar system are explained below.

Solar Power Panel

The main component of a DIY solar system is the solar panel. Each panel is made up of several individual photovoltaic cells. A photovoltaic cell is a small device that converts sunlight into electricity through the photovoltaic effect. These cells are also called solar cells. Since individual cells only produce a relatively small amount of electricity, they need to be connected to other cells to achieve practical power levels. Solar cells are mechanically mounted on a stable substrate and then wired together to obtain the necessary amount of energy. The substrate is mounted in a frame that has a weatherproof glass or plexiglass cover.

Then several solar modules are connected to a bracket to form an array of solar panels. The array can be attached to a roof, a pole, or even close to the floor. The position of the array must be in a position that receives the most direct sunlight with a minimum of shadow.

Solar-Powered Battery

Batteries store excess energy from solar panels available at night or when sunlight is not available. Batteries are electrochemical devices. Their performance depends on the climate, temperature, charge/discharge cycle, and age. Although several battery technologies exist, lead-acid batteries offer the best performance per dollar and are the most commonly used type in solar power systems.

The limit of a battery is given in amperes hours at a specific voltage. A 100-hour, 12-volt battery would deliver 12 volts to 1 ampere for 100 hours. A bank of batteries must be sized so that electricity can be kept for five days without sunlight. The battery type must be a deep cycle type. This means that the battery can drain most of its capacity before charging. A car battery is a flat cycle type and, therefore, not suitable for solar energy applications.

Solar Charge Controller

Usually, a controller is needed to extend the life of the battery bank. The main function of a controller is to prevent the battery overcharging. A controller monitors the battery voltage and reduces or stops the charging current when the voltage increases. When the battery voltage drops, the controller increases the current to charge the battery. Controllers are evaluated based on the amount of electricity they can process. We recommend installing a larger capacity controller first if you plan to expand the system in the future. A larger controller usually doesn't cost much more.

Solar Power Inverter

The battery bank stores direct current from the solar modules. An inverter converts a low DC voltage from the batteries to a higher AC voltage, e.g. 120 or 240 volts, which can supply electrical devices. The inverter also sets the AC frequency, which is 60 Hertz (cycles per second) for the United States. It also offers a sinusoidal waveform, necessary for most electronic devices. The inverter must be designed for the required system performance. The brand inverters are very reliable and have a sales efficiency of around 90%.

INSTALLATION OF A SOLAR PANEL ON-BOARD YOUR BOAT

After a five-day trip aboard our fishing boat, we often had to start the generator, set to power the 115 volt AC freezer and maintain temperatures.

Solar panels have been used successfully since the mid-1950s and were originally used for manned space research. Costs have gone down since around 2004 when their popularity has increased. And now that the green pressure continues, solar panels are being accepted like never before. You can find many retailers who sell solar panels on the internet. However, I can't find a detailed description of how to determine what to buy and how to install it. So, this piece was written during the process and as such, is a true learning article.

What Exactly is a Solar Panel, and How do they Work?

Solar collectors are all modules that use the sun's thermal energy to generate electricity. A solar module can be referred to as a photovoltaic module, the name used in the industry for modules designed to generate electricity from the sun's emissions.

Despite the discussed group of solar modules, almost all solar modules are flat. This is because the panel surface must be tilted at an angle of 90 degrees from the sun's rays to allow the best angle to absorb the sun's rays. Solar panels can absorb energy from the sun through the accumulation of solar cells on their surface. Similar to how a plant can absorb energy from the sun for photosynthesis, solar cells function comparably. When the sun's rays hit the solar cells on a photovoltaic module, energy is transferred to a silicon semiconductor. The energy is then converted into a direct current and then passed through the connection cables to finally enter a storage battery.

Types of Solar Panels

The types of panels that are generally used in yacht applications have multi-crystalline or amorphous thin-film cells. Multicrystalline sheets are the oldest and strongest technology available. With the correct size and combination with suitable batteries, these are the panels that are suitable for the operation of large loads such as. B. cooling can be used.

Thin-film amorphous solar modules are only about 50% as effective multi-crystalline modules, but they can be purchased in flexible versions so that they can roll or fold or take the shape of a boat cabin. Often, they do not have sufficient yield for a substantial supply of energy, but they can be used to easily charge a bank of batteries.

How much Electricity can Solar Cells Produce?

We usually measure solar panels by power and buy them this way. You can get solar panels for boats from 10 watts up to 200 watts or even more. However, it is easier to understand if we change watts to current. We calculate these values by multiplying the number of hours the panel is in full sun (usually defined as 5 per day in Florida) by the power of the panel. For a 195 watt solar panel, the quantity produced would be 195 x 5 hours = 975 watts per day. We can therefore calculate 975 watts/12 volts = 81.25 amps per day.

Energy Consumption

Before you start thinking about the dimensions to buy for your boat, you will need to compile an energy budget to determine which sources of energy consumption you have at anchor. I'm using the scenario at anchor because that's where you use the most energy.

For example, if you have three internal lights, each with two amps and leaving them on for 4 hours per night, the consumption will be 3 x 2 x 4 = 24 AH/day. We are not worried about chases and electronics since they are unlikely to anchor.

DC Loads - Calculate how many Amp-hours each Device Consumes

- Home lighting
- Anchor lights
- cooling down
- freezer
- Electric toilet
- Freshwater pump
- Sanitary system
- stereo
- other

Inverter loads

Inverter loads also use DC, but they power AC devices and equipment. If you want to change the watts in amperes, use (12 watts/12 volts = 1 amp).

Calculate the amp hours used by each device

- computer
- microwave
- cooling down
- freezer
- heating
- Hairdryer
- TV
- other

Add the total daily energy consumption AH/per day

Solar power generation

Different power sources similar to solar modules can replace the amperes/hours taken from the batteries. Similar to the electricity budget that determined your use, you also need to calculate the next delivery of amperes hours. Follow the formula - (12 watts/12 volts = 1 ampere). But be careful, the formula is only a yardstick: full accuracy can only be achieved if panel production is constant and a solar panel sometimes works ineffectively due to the cloudy sky. The daily electricity consumption in AH/day for the generation of solar energy is displayed. Your solar energy production must be greater than consumption. If not, select a larger power and recalculate it. Buy more solar modules than you think is necessary. Some professionals recommend at least 30% more.

A typical example: 100 watt / 12 volt solar panel = 8.3 amp x 5 hours = 41.66 AH / daily production

Install your Solar Panel

Where do you put your solar panel on-board your yacht after receiving it? As already mentioned, it is better to install the panel at 90 degrees from the sun. This way, you get the best energy generation. But on yachts, it is at best difficult to find a suitable place. Many boaters have placed them on clamps fixed to the rails, and I have seen them positioned on the white uprights of the trawler. But wherever you choose to mount them, keep in mind that, to get the best out of them, they should be outdoors, away from any boom shadows, ship's radar bows, or cabin structures. Remember that due to the tides, the boat at anchor turns to the sun twice a day. I decided to mount our panel horizontally on the top of the aft deck of the boat. Here it has the best view of the sun and is clear from the shadow of the radar arch when the boat swings at anchor. The inclination towards the sun is not exactly ninety degrees, but it must be sufficient. I chose a 195-watt panel, so I have almost 50% reserve in my panel to compensate for the low inefficiency of the sun's angle. We received the panel from 'Sun Electronics' in Miami, sunelec.com, because it had the best prices I could find on the internet. However, keep in mind that panels must be shipped with goods as they are heavily packaged to reduce the risk of damage. Therefore, be sure to calculate these costs when purchasing.

What is the Best Way to Attach the Panel to your Vessel?

There are numerous manufacturers of rails and mounting brackets for solar panels, but almost all of them are designed for roof or floor mounting. West Marine has a product for mounting small panels on rails. A good number of boaters make their supports. I found support developed by Sunsei, known as the Sunsei glue assembly kit, which is attached to the ship and the plate with the 3M 5200 marine adhesive. The bracket allows the panel to be installed at a distance of about 5cm below the panel for ventilation. I also didn't have to drill holes in the hardtop. You can find these brackets on 'amazon.com.'

Panel Wiring

The electrical wiring of ships is incredibly specialized and dangerous. If you are unfamiliar with this part of the activity, please contact an experienced marine electrician. Your panel is pre-wired for connection to your ship. However, you must supply the connection cables also sold by your panel supplier. They are called MC4 cables. The cables are produced in different lengths to meet your needs, with a plug and socket connection.

You also need a controller. The controller regulates the current from the panel to the batteries and keeps the batteries charged without interrupting the charging process. Many controllers are simple, others have LED indicators showing the amount of charge, etc. The more unusual the functions, the more expensive it becomes. Your panel dealer can suggest a controller that meets your needs. I chose a controller developed by Specialty Concepts. It's simple, but it does the job. The company staff will help you choose the best model for your panel. When you contact them, they want to know what size (power) you are needing and what voltage is required. Check out specialtyconcepts.com. I also received my controller from 'Sun Electronics' employees in Miami.

'Special Concepts' employees also calculated how heat affects the flow of electricity and suggests that their controllers should not be installed in engine rooms, as the heat generated reduces the effectiveness of the controller by approximately 25%. I put mine in the control panel under the lower helm. Finally, the right size cables are needed to get from the controller to the batteries and a fuse to connect the controller to the battery bank. When choosing the right fuse, you need to determine the short circuit current for your panel and classify the breaker with 125% of this number. This gives you the amperage of the switch you need. You can also find information on this in the instructions for the use of the controller.

Having a solar panel to save your batteries may seem like a great plan, but you want a way to keep an eye on your batteries. I also decided to install a Trimetric 2025RV battery monitor from 'bogartengineering.com.' This smart device is connected to the battery bank to allow a real measurement of the voltage introduced into the bank, the amplifiers used by the yacht, the percentage of a full charge on the bank, and the ampere-hours used since the last charge.

Panel Functionality

This is why we have now installed a 195-watt solar panel as well as an 1800 watt inverter and a battery bank with 443 amps hours. This week we ran our hook tests with clear skies. The freezer is the main consumer at 60 amp hours, followed by the refrigerator. Did you know that a standard anchor light consumes 18 amps every night? I think now I'm going to check the LED lamps.

The battery monitor indicated that our actual consumption from the battery bank was only 65 amps, which indicates that the rest was obtained from sunlight, 42% from the sun. Now we turn on the unit for about an hour to fully charge the battery bank.

SOLAR PANELS FOR BOATS

The era of solar panels is clearly upon us. You see them regularly on stop signs, on the roofs of homes, and in countless other daily operations. But let me tell you about the next wave of vehicles that use solar panels for boats.

Whether you are an avid sailor or just an occasional fisherman who owns a boat, you have already thought about the fuel costs and the many problems that can arise from the constant addition of petrol to your engine, and the problems, which can arise from the mixing of water and petrol. Engines aren't cheap, are they?

We certainly don't park our boats in our garages at night. If they are just floating on the water waiting for us to use them, how nice would it be if they worked for us during these idle periods?

With the new designs, the artists have incorporated a combination of solar-powered cells with a kite-like device that opens up and allows for high speeds and an easy start.

This is not to say that they just attached a sail to a boat, threw it on a couple of solar panels, and called it a day; far from it. There are moving parts, start and stop mechanisms and an engine type.

The plans were not outsourced to people who know nothing about sailing; In fact, the artists themselves would not have been able to develop the designs if they had not had one foot in the water for a lifetime.

The new models are called Kitano, designed by Stefanie Kracke and Volitan, and also known as 'Flying Fish.'

Look for these new models that will soon be made available to the public. As with all new inventions, in particular, the new energy-efficient models provide a price that only the elite can initially afford. But as more and more are sold, prices will drop. Given the current trend, Joe Smith is expected to buy one for his boat family by 2020.

Another step in the right direction for sailing companions. If we as a community support these products and continue to encourage yacht artists and designers, there is no reason why this cannot become a reality for us.

DIY SOLAR POWER GENERATION TIPS FOR YOUR BOAT

Electricity at anchor is almost a must for most sailors or motor sailors, especially if you are traveling with your family. Electricity is needed if the admiral is to use the hairdryer after a hot shower, if hot water is needed for dishes or if you just need to keep the refrigerator cold. A marine solar panel system can generate this electricity silently and efficiently for you.

There are other ways to generate energy; operate the engine (s) or run an on-board or portable generator. Both methods require energy consumption to generate energy. This takes the form of burning fossil fuels. What if I could get the energy for free and then convert it to usable AC / DC electricity? With a system of marine solar panels, you can, and as a bonus, you can design and build the system yourself if you have the minimum mechanical suitability.

Don't get me wrong now, there are a variety of pre-packaged marine solar module kits currently on the market, and the selection seems to grow day by day. Check your local equipment or even search online. Depending on the size of the panels, you will find numerous kits, the price of which is up to $700 and above. However, with a few free weekends and some of the mechanical skills mentioned above, you can design, build, and install your solar system for a few hundred dollars. Remember that even if you buy a 'standard' system, modifications will most likely be necessary to make it usable for your boat.

The first thing to bear in mind with a DIY project is to determine what is needed to complete the activity. In this case, additional precautionary measures should be taken, as your finished solar panel system must be suitable for harsh marine environments.

First, you need to determine the overall performance required for your travel style. Add the total energy consumption (in amperes), you would need add the anchor and divide the total current by 2. This gives you the power you need.

The type of panels to be integrated into the project should also be considered. For example, monocrystalline photovoltaic cells are highly efficient electric generators that take up little space but lose efficiency in low light or indirect solar conditions. Polycrystalline cells take up even less space, but still, lose efficiency in low light conditions. The other possibility would be the use of amorphous photovoltaic modules, which have a lower overall efficiency with a larger area at the same time but do not lose efficiency in low light conditions.

Regardless of which direction you choose, keep in mind that there should be at least a general understanding of how electricity is generated with a marine solar panel system. My advice is to research and carefully plan your project to avoid mistakes. When you finish the project and step back to admire your work, you'll not only know the sense of self-actualization, but also that you're about to save money (to buy more fuel or drinks).

SOLAR ENERGY FOR YOUR MOTORHOME

Many motorhome owners believe that their vehicle is as much a home as their house. For this reason, it only makes sense to consider a solar-powered motorhome and a solar house, especially if you spend a lot of time on the road. And like everything that has to do with solar energy, there are not only advantages for the environment, but also solar energy is valid in a practical sense.

For example, solar systems weigh less than a small generator that campers are usually equipped with, not to mention the lack of petrol they need. And of course, camper owners want their car to run faster and better. So, less weight (and smoke) is always the best choice! A luxury comfort offered by solar systems is that they are silent. Campers are generally designed to escape everyday life and relax. Solar systems require no special operational skills, no assembly, practically no maintenance, and are excellent for the environment! You can feel good knowing that your family vacation or your trip will not pollute the beautiful scenery you bought for your camper.

Now that we know why solar power is an excellent alternative to traditional recreational vehicle power methods, how can we apply this knowledge to our motorhomes? Below are some basic things to consider to create an efficient solar system for your travels.

First, determine the size of your cargo. The factors to consider are how many people will travel with you, how big your particular motorhome model is, and for what activities you are likely to use your motorhome. For example, a rough estimate of the electricity needed for a family campsite is 325-watt hours. A watt-hour is the amount of energy consumed by a load of one watt that consumes electricity for one hour. These include light bulbs, heating fans, water pumps, televisions, radios, coffee makers, and the estimated useful life of these items per day. For dry camping enthusiasts, solar energy is a must. Dry camping includes numerous camping trips with limited services. Your solar power supplier can help you determine what your motor home needs.

How much sunlight do you expect in your location? It is important to find out how much estimated sunlight your area gets when you travel, and there are many resources available to help you acquire this knowledge. The Center for Renewable Energy and Sustainable Technology (CREST) can help you find out how much sunlight comes to different places in different seasons. This also applies to the National Renewable Energy Laboratory (NREL). The solar panel can also be adjusted to various angles during the day to absorb the maximum amount of sunlight.

ALTERNATIVE POWER SUPPLY FOR CAMPERS

The campground trip means you've brought tents and nothing else and this is how camping should be. However, today we have motorhomes to allow us to bring technology with us. But if you're wanting to spend time actually camping, you don't need devices like televisions or CD / DVD players, computers, netbooks, etc. They are not necessary and reduce the quality time in the forest. If you want to see something exciting on TV over the weekend, stay at your house and go fishing next weekend. Camping means to separate from these devices. Camping is listening to wolf calls at night, wandering in the forest during the day, taking photos, and leaving only footprints.

Motorhomes have pre-installed energy sources, and some used motorhomes or caravans also include generators. But as far as possible it is necessary to reduce electricity consumption, except when it is very hot or cold outside. Here are some of the accessible energy sources for campers and motorhomes for sale.

- Motorhomes have V8 and V10 engines that leave the air conditioning on while the engine is running. At the same time, connected power (110-230V), propane, or fuel-powered generators can be used to power the refrigerators. Microwave ovens, gas stoves with 2-3 burners, and the rest of the services.
- Usually there are two batteries in campers and motor homes. You have the vehicle battery or the SLI (start, lighting, ignition) and house batteries. The home battery is used to power devices and lights when the engine is turned off.
- Propane gas, or more commonly known as liquid propane (LP), is used to power heaters, 3-4 burner stoves, grills, water heaters, or to operate refrigerators. There are two types of LP tank containers, the Department of Transportation (DOT) and the American Society of Mechanical Engineers (ASME) containers. The former is often found in smaller motorhomes such as truck trailers and pop-up motorhomes. At the same time, the latter is particularly preferred for motorhomes that are mounted horizontally.

Power generators are generally used to power air conditioners in campers. They are also needed to charge the home battery. The generators are measured in watts, and the smaller campers need

3000 watts to supply electricity to the air conditioning systems. In comparison, the large campers need generators with a power of 6000 watts to power different air conditioning systems for the whole area. But they stifle the blissful sound of the forest with the sound of their engines. Alternatives to power generators are solar power systems for campers, they don't require much maintenance, aren't noisy, and are less expensive to use. Whether you camp from time to time or every weekend, solar power is worth considering. However, it is also necessary to contemplate the ability of the solar energy system to supply precisely the vital energy, which is limited depending on these things:

1. The number of solar modules installed on your RVs

2. Or the amount of solar energy you extracted from the sun.

a) The former depended heavily on how long you want to spend camping. To reduce complicated calculations of how many solar panels are needed to operate your motorhome or camper van. You have to take into account that a 200 AH home battery needs at least 75 and 120 watts to work well. Follow these guidelines and calculate how much you need. Or you can simply buy two solar modules, which is a safe number, and add more modules if needed.

b) Some brands of solar modules do not work well if one of the modules does not receive the necessary sunlight during the day or in the shade of the sun. However, there are some brands specially designed to skip the shaded area so as not to interrupt the energy acquisition process. You need to choose the brand carefully to avoid losing the power supplies if you get them wrong.

Some of the best solar power producers offer to install solar panels and to provide free or minimal payment for the conversion or process of electricity. Have the product installed on a camper properly if possible to avoid electrical problems due to incorrect installation and, among other things, to guarantee 100% performance.

THINGS TO KNOW ABOUT RV SOLAR POWER

For nature lovers, there is nothing better than taking off in a camper and exploring the landscape. Unfortunately, these large, bulky vehicles also require a lot of off-road energy. We do not yet have an efficient solar-powered engine on the market, and the few hybrid motorhomes that have been developed are still in the concept stage. However, numerous solar power kits can be used for recreational vehicles, reducing or eliminating the need for expensive electrical connections on campsites. There are a few things to consider when planning to install such a system.

If you choose the right solar kit, you must first collect some information. The roof area should be available for installing a solar system. So, approximately how much direct sunlight will be available on average, both while driving and when parking in the sun? Finally, it should be determined how much electricity is used and how many devices need to be powered by electricity at any given time.

Taking these facts into account, it becomes possible to make a fair decision about the number of solar modules needed to create an array, what type of photovoltaic modules to use, and the amount of memory necessary for the battery. There are three types of panels that are commonly available on the market. The cheapest is made of amorphous silicon. However, it is the least efficient at converting light into electricity. The next step is polycrystalline silicon, which is more expensive but more efficient. The slightly more costly monocrystalline cells are the best with the highest available efficiency. The only problem with monocrystalline cells is that they cannot be cut into waffle-like square pieces and, therefore, cannot be packed together as cleanly as other cell materials. Monocrystalline is the best choice if there is enough roof space. It may be advisable to consult a professional when making this decision and when you are ready to install it.

In some cases, when there is not enough space to install a sufficiently large array, it may be necessary to use a gas-powered generator to integrate the generation when the solar system is overwhelmed. As with any complicated installation, a specialist should always be consulted here.

When the energy prices go up, I pay the camping fees for electricity. Eliminating these costs can go a long way in balancing gasoline costs and repressing the unrest that can result from a motorhome's huge carbon footprint.

DIY SOLAR MOTORHOME SYSTEMS - HOW CAN I CREATE A DIY SOLAR MOTORHOME KIT?

Why should a motorhome owner want a do-it-yourself solar system? One of the reasons is the high cost of a customized solar energy configuration. A motor home usually has two ways of generating electricity for the passenger compartment. The first option is to use a gasoline generator to generate electricity when the vehicle is parked. Some people don't like this option because it is noisy and causes a lot of pollution. It also has high operating costs because it is necessary to continually fill it with petrol.

The second option, which is more desirable, is to use solar panels to generate electricity. It is clean, efficient, non-polluting, and requires minimal maintenance—the only negative aspect is the installation cost. A customized solar panel for motorhomes that can produce up to 440 watts can cost up to $3000! Many motorhome owners prefer to spend the money on other improvements. Due to the high cost, many motorhome owners are turning to the DIY solar kit for motorhomes. The system is not perfect, but it can produce useful electricity for the home at a lower cost.

A homemade solar system is quite simple. The main components are solar panels, the charge controller, battery storage, and the inverter. The other various parts that support the system are ammeters, voltmeters, automatic switches, and safety barriers. The most important part of the whole system is that solar panels, must be installed on the roof of the vehicle. If your motorhome has a flat roof, this is a simple process. However, if the vehicle roof is curved, a special process is required to take this into account.

Assembly plays an essential role in ensuring that photovoltaic (PV) panels operate at their optimal capacity. If placed flat on the roof, the panels do not have the best angle for capturing sunlight. A mobile bracket that can be adjusted or tilted is the best solution.

All photovoltaic modules must be connected to the charge controller, which uses the electricity generated to charge the battery bank. When building the battery bank, it is best to choose a real deep cycle battery. This is the type of battery used in electric golf carts. They are more resistant and can be unloaded to a lower level without damaging the internal panels. The final step is to connect an inverter to the battery and then pass an extension around the living space of the camper. Thanks to the sun, you can now enjoy free electricity on the open road.

THE FOUR ESSENTIAL COMPONENTS OF A SOLAR MOTORHOME SYSTEM AND WHAT THEY DO

The order of the elements in a solar system is:

1. Solar panels

2. Charge regulator

3. Batteries

4. Inverter

The installation of a solar system for lights and on-board devices is a simple process in which batteries are charged with solar energy. Each component has its task, and when it works together, you have your power plant with you.

Solar panels are flat, rectangular modules that are usually mounted on the roof of a motorhome. They convert solar energy into electricity and send it to the charge controller via cable.

The charge controller essentially maintains the correct charge voltage supplied to the battery bank. When the sunlight is brighter, the panels generate more voltage. As the voltage increases, the controller adjusts the charge to prevent the batteries from being damaged by overcharging or excessive voltage.

Electricity is stored in batteries. Golf cart batteries are widely used in motorhomes due to their deep cycle, which means they can be discharged to a lower level and charged many times without damage. They can be downloaded up to 80% before being reloaded. The rule of thumb, however, is that the depth of discharge before charging is between 20% and 60% for best performance. 'Golf cart battery' is a term commonly used to describe most 6 or 8-volt deep cycle batteries. They can be wired in series with a series of 2, 4, or 6 batteries to produce 12, 24, or 48 volts.

There are three different types of batteries.

- Flooded batteries are the cheapest. They are lead acids with caps to add water. They emit gas during charging; therefore, they cannot be used indoors and must be located in a ventilated area. They are inexpensive, work well, and can last for many years.

- The second type of battery is the sealed gel battery. They do not release gas during charging; therefore, they can be used in closed rooms without ventilation. This makes the difference because the batteries have a relatively stable temperature inside, which allows them to function better. They are more expensive than flooded batteries but still relatively cheap.
- The third type of battery is the Absorbed Glass Mat. Many consider AGM batteries to be the best for solar energy. They are leak proof and do not emit gas during charging. They offer superior performance and are also the most expensive. When choosing the type of battery for your motorhome, your budget will probably be the most critical factor, as they all provide excellent service over the years.

The fourth component of the solar system is the inverter, which converts the low voltage direct current generated by the solar modules into the 120-volt alternating current used by most lights and on-board devices. There are three different types of inverters.

- The cheapest inverter is a square wave. They are not very efficient and can damage many sensitive devices such as

computers or HD televisions.
- Modified sine wave inverters are also cheap and work somewhat better than a square wave. However, they can also damage high-end devices. Devices that use motor speed controllers or timers may not work correctly with this type of drive.
- Pure sine wave inverters provide your devices with pure, clean electricity that corresponds at least to the energy supplied by your energy supply company. Pure sine wave inverters are more expensive, but they can safely operate almost all types of AC devices. They are usually computer-controlled to automatically turn on and off when AC loads fluctuate. Most devices work more efficiently and consume less energy with a pure sine wave inverter.

You can also add a monitor located inside the coach. Monitors have different functions, and you can choose the one that best suits your needs. With the monitor, you can keep track of the state of charge (SOC) of the batteries. It shows how full they are, how fast they load, and how fast they are unloaded.

The panels convert sunlight into electricity and pass the cable to the controller, which keeps the charge at the right voltage and quantity. Electricity is then fed into batteries for storage. When the light switch is turned on, the inverter switches it from direct current to alternating current at 120 volts, and the light turns on. It is a fairly simple system, and once installed, you can save a lot of headaches and money.

Before you get started, you should take a look at the equipment so you can see exactly what you are working with.

REQUEST SOLAR PRE-WIRE

If you buy a new camper, indicate to the manufacturer and dealer of the camper that you want to install the solar pre-wire at the factory! This option costs very little if it is performed before the installation of the walls, so you can save hundreds on your solar system with this simple step! Also, be sure to use a high-quality solar pre-wire kit of at least 10mm thickness. Some manufacturers opt for the cheaper Prewire kit, which only supports a maximum of 200 watts of solar energy and is not close enough for most campers.

OUR SOLAR INSTALLATION QUESTIONS - KNOWLEDGE AND AVOIDANCE OF THINGS

With our latest electric, solar kit, we have all encountered some installation problems. I will list them from the worst to the not so terrible:

Assembly parts: The installer used rivets on our roof, which turned out to be an error. Some rivets came loose, and we had to replace all of the rivets with huge wood screws. Ask your motorhome manufacturer about the roof material and the type of anchorage that can best be kept on the roof. Even the motor home company may not be 100% sure. Therefore, contact the manufacturer of solar modules for more information (customer service and warranty are the reason why it is essential to choose a reliable solar company).

Correct MPPT placement: Do not install the MPPT controller upside down. When fully loaded, the fan turns on to dissipate the heat generated in the solar controller. When the fan starts, heat is discharged, but when the heat rises, the MPPT controller does not cool quickly, which can expose the solar controller to overheating.

Length of the thread: The threads must be cut short, but not too short! One of our solar connection cables is rather rigid and is pulled slightly when the panels are tilted. This extra pressure on the link can mess with your connection.

THINGS TO ASK BEFORE ANYONE WORKS ON YOUR RV:

Have you perused the headings?

If the kit was supplied with the instructions, read them until you understand all aspects of the installation. You shouldn't rely on your RV technology to understand the pros and cons of your solar kit. While the technician may have installed solar panels in the past, he may not have installed a system like yours. If possible, try to be available and in the shop to confirm that everything is done correctly.

Does your solar kit contain everything?

Make sure that the kit includes all mounting hardware such as screws, brackets, and solar cable connectors for the panels.

Sealants and adhesives are generally not included (due to expiration dates). Therefore, confirm with your installer that they have stock available and that they have not expired.

Where are the panels installed?

Measure your roof and panels and then draw your favorite solar panel configuration. You may want to draw two versions in case the installer can't layout your favorite solar panel configuration (you never know what's right under the roof, so it's good to have a backup plan ready).

Once you have an idea of the arrangement of the panels, place the boxes on the roof in the same method to make sure everything fits as expected before you start transporting the panels. To be more specific, the lower half of the box can be cut according to the size of the solar module while keeping the top of the box to cover the modules during installation.

Where are the shadows possible?

Keep panels as far away as possible from rooftop air conditioners, fan covers, or satellite / TV antennas. Remember that the sun is lower on the horizon in winter, and the shadows become longer. As a result, the air conditioning may not cast a shadow at midday in the summer but maybe wholly shaded in the winter.

What size do I need for solar and battery cables (wire thickness/size)?

The length of the solar cable must not exceed 25 feet from the solar system to the battery bank. This diminishes the 'voltage drop' (loss of charging power). If the link is excessively long, you will lose a portion of the sun-based vitality before it arrives at the batteries.

The sun-based link must not be under 10 meters when all is said in done, the bigger and thicker the wire, the better the force transmitted through the link.

An 80 amp electrical switch is associated with the sun oriented positive because of the high voltage. No one can tell when the board may be turned off.

The Big Takeaway: The thicker and shorter the wire, the less power is lost.

Wire the solar modules in a series or parallel?

We decided to connect the modules in a series based on the number of solar modules, the size of our wires, and the MPPT solar controller. This is why:

Since we have connected all six panels in series, 120 volts DC are output. Due to this high voltage, we can switch with 960 watts on a wire with an AWG value of 10 from the solar modules to the solar controller. This effectively reduces the current flowing through the cable to around 9 A. Our MPPT Outback controller lowers the voltage from 120 volts to 12 volts to charge the battery bank. In the wake of leaving the MPPT controller, the current is around 55 amps. For the solar controller, we used an AWG cable no.2, being short and resistant to handle this large amount of energy.

What is the difference between the connector types (MC4 vs. SAE)?

MC4 connectors can usually manage 30 amps, including solar cable no.10, which are waterproof and cannot be easily separated from the guide vibrations. Typical SAE connectors are designed for 20 amps and are less robust and, therefore, cheaper.

Best practices for installing a waterproof cable?

Even if the camper does not have a solar pre-cable, a cable entry plate can be used to position the solar modules as close as possible to the battery bank. The cable entry plate we installed has a bottom channel to ensure that the sealant is 100% waterproof at the edges when used correctly.

Do I need a remote for my inverter or solar controller?

As soon as the settings have been entered on the inverter and the solar charge controller, we generally configure and forget them. However, we use our remote controls daily to monitor battery power, to see how much solar energy we are feeding and how much energy we are using. The remote controls for the solar controller and the inverter should be easily accessible in the living areas or from the control panel of the camper (with tank monitor, air conditioners, etc). When it's sunny outside, check these remote controls often and celebrate when the battery is 100% charged while you're in the middle of nowhere!

Where should I install the components?

Inverters, solar controllers, and batteries must be kept away from water. The batteries, in particular, must be in a separate and enclosed area with ventilation (ventilation is a must for lead-acid batteries, but check with AGM or lithium batteries).

Do not place inverters or MPPT charge controllers in the motor home, as they often generate heat and not only heat the motor home in the summer, but the fans can also be noisy. We recommend installing these items in a watertight closet.

The fuses for the inverters must be as close as possible to the battery bank. All fuses should be easily accessible when they need to be replaced.

Is it safe to go on the roof with the solar panels?

You have to climb on the roof to clean the panels, tilt them and check the wear of the seals. It is essential to leave a path to walk safely around the roof. You don't want to walk on solar panels. To avoid tripping over less exposed cables on the roof, you should be able to route most of the cables under the panels so that there are as few cables exposed as possible.

HOW SOLAR ENERGY WORKS: ON-GRID, OFF-GRID AND HYBRID SYSTEMS

All solar power systems operate on the same basic principles. Sun-powered boards support sun-based vitality or daylight into direct current, utilizing the supposed photovoltaic (PV) impact. The direct current would then be able to be put away in a battery or changed over into substituting current by sun oriented inverter, with which family unit apparatuses can be worked. Depending upon the kind of framework, abundance sun oriented vitality can likewise be taken care of into the power lattice to give advances and further decrease your power costs.

The three main types of solar energy systems:

1. On-Grid - also known as the solar system connected to the grid or powered by the grid

2. Off-grid - also known as autonomous power system (SAPS)

3. Hybrid - Solar plus battery accumulator with mains connection

We will first describe the common components used by all three types before delving into the different systems and how they work.

Main components of a solar system:

- Solar panels
- Solar panel mono PERC Winaico 300W
- Solar panel mono PERC Winaico 300W

The most modern solar modules are made up of many silicon-based photovoltaic or photovoltaic cells, which generate direct current from solar energy. The individual photovoltaic cells are connected within the solar panel and connected to neighboring panels via DC cables. Note: It is light energy or radiation, not heat, that generates electricity in photovoltaic cells. Solar modules, also known as solar modules, are installed together in 'strings' to create a so-called solar array. The measure of sunlight-based vitality produced relies upon various elements, including the direction and point of the sun based boards, the efficiency of the solar panel, and any losses due to shading, dirt, and even ambient temperatures. There are many different solar module manufacturers on the market. So, it is worth knowing which solar modules are best and why.

Solar panels can generate energy in cloudy weather. However, the amount of energy depends on the thickness and height of the clouds, which determine the amount of diffused light that can be transmitted. The measure of light vitality is known as sun introduction and is usually referred to as Peak Sun Hours (PSH) throughout the day. The amount of PSH or the average hours of sunshine can vary greatly depending on the location during the year.

Solar Inverter

Solar panels generate a direct current which must be converted into alternating current for use in our homes and businesses. This is the main task of the solar inverter. In a string inverter system, the solar panels are connected in a series, and the current is directed to a single inverter, which converts the current into an alternating current. In a microinverter system, each panel has its micro-inverter, which is attached to the back of the panel. The panel still produces a direct current but is converted into the alternating current on the roof and fed directly to the electrical panel.

There are likewise, further developed string inverter frameworks that use little force analyzers associated with the rear of each sun-powered module. Performance optimizers can monitor and control each panel individually, ensuring that each panel operates at maximum efficiency in all conditions.

Battery

Solar energy batteries are available in two main types; lead-acid (AGM and Gel) and lithium-ion. There are many other types, such as redox and sodium-ion batteries, but we will focus on the two most common. Most modern energy storage systems use rechargeable lithium-ion batteries and are available in many shapes and sizes that can be configured in different ways.

The measured battery capacity in ampere-hours (Ah) for lead-acid or kilowatt-hour (kWh) for a lithium-ion. However, not all capacity is available. Lithium-ion batteries can typically supply up to 90% of their available capacity per day. In comparison, lead-acid batteries should only supply 30 to 40% of their total capacity per day, extending battery life. Lead-acid batteries can be discharged entirely, but this should only be done in an emergency.

Off-grid solar systems require special off-grid inverters and battery systems large enough to store energy for two or more days. The systems connected to the hybrid network use cheaper hybrid inverters (battery inverters) and require only a battery large enough to supply energy from 5 to 10 hours depending on the application.

Power panel

In a standard solar system connected to the grid, alternating current is led from the solar inverter to the control panel, where it is routed to the various circuits and devices in the house. This is called 'grid measurement,' in which excess electricity generated by the solar system is fed into the grid via an energy meter or a battery storage system if you have a hybrid system. However, some countries use gross metering, in which all solar energy is exported to the electricity grid.

Hybrid systems can both export excess electricity and store excess energy in a battery. Some hybrid inverters can also be connected to a dedicated backup panel capable of powering some 'essential circuits' or critical loads during a network or power outage.

1. System on grid

Grid or grid solar power systems are by far the most used and widely used by families and businesses. These systems do not require batteries and use conventional solar inverters and are connected to the public electricity grid. Any excess solar energy generated is exported to the network, and usually, an input tariff (FiT) or credits for the energy that is exported is obtained.

Unlike hybrid systems, grid-connected solar power systems cannot operate or generate electricity during a power outage for safety reasons and because power outages usually occur when the power grid is damaged; if the solar inverter still supplies electricity to a damaged grid, this would endanger the safety of the people who repair the grid faults. Most hybrid solar power systems with battery storage can automatically disconnect from the grid (the so-called island formation) and continue to supply electricity during a power outage.

If necessary, the batteries can be installed in networked systems later. Tesla Powerwall 2 is a popular AC battery system that can be added to an existing solar power system.

In a network system, this happens after power has reached the panel:

The measuring device shows when excess solar energy flows through the meter, which calculates how much electricity is being exported or imported (purchased).

Measurement systems work differently in different countries. In this description, I assume that the meter only measures electricity that is exported to the grid, as is the case in most of Australia. In some states, meters measure the total solar energy generated by the system. Therefore, your current flows through your meter before reaching the panel, not after. In some areas (currently in California), the meter measures both production and export, and consumers are charged (or credited) for the net electricity consumed over one month or year. I will explain more about measurement in a later blog.

The electricity that is fed into the grid by your solar system can, be used by other users of the grid (your neighbors). If your solar system is not in operation or you consume more electricity than that produced by your system, you will import or consume electricity from the grid.

2. Off-grid system

An off-grid system is not connected to the mains and therefore requires battery conservation. Off-grid solar power systems must be designed to generate enough electricity all year round and have enough battery capacity to meet the needs of the home, even in winter, when there is typically much less sunlight.

Due to the high cost of batteries and off-grid inverters, off-grid systems are much more expensive and are typically only needed in more remote areas that are far from the grid. However, battery costs are decreasing rapidly, so there is now a growing market for off-grid solar battery systems in cities.

Several types of off-grid systems will be covered in more detail later, but for the moment, I will make it simple. This description applies to an AC coupled system. In a DC-coupled system, energy is first sent to the battery bank and then to the devices. To learn more about how to build an efficient off-grid house, visit our affiliate site, which is off-grid.

Simple, inexpensive, small, DC coupled, off-grid solar power systems use solar charge controllers to manage battery charge and a simple inverter to provide AC power.

There is no public power grid in a grid-independent system. As soon as the home devices consume solar energy, excess electricity is sent to the battery bank. Once the battery is full, it is no longer powered by the solar power system. When the solar system is not working (at night or cloudy days), the devices consume energy from the batteries.

In times of the year when the batteries are low, and the weather is very cloudy, an emergency power source is usually needed, e.g. an emergency generator or a generator. The size of the generator set (measured in kVA) should be sufficient to power your home and, at the same time, charge the batteries.

3. Hybrid system

Modern hybrid systems combine solar and battery storage in one and are now available in many different shapes and configurations. Due to the reduction in battery storage costs, systems that are already connected to the mains can also use battery storage. This means that solar energy generated during the day can be stored and used at night. When the stored energy is depleted, the network acts as a backup so that consumers can enjoy the best of both worlds. Hybrid systems can also charge batteries with low-cost electricity during off-peak hours (usually after midnight until 6 am).

There are also several ways to design hybrid systems, but for now, we'll keep it simple. You can find more information on the various hybrid and off-grid power systems in our detailed guide to solar battery systems for private homes.

In a hybrid system, excess electricity is sent to the battery bank as soon as solar energy is used by the devices you own. As soon as the battery is fully charged, it no longer receives electricity from the solar power system. Battery energy can then be discharged and used to power your home, usually during the peak of the evening, when electricity costs are generally the highest.

Depending on how your hybrid system is installed and if your energy supplier allows it, the excess solar energy that is not required by your devices can be exported to the network via your meter after it has been fully charged. When your solar system isn't working, and you've run out of usable electricity in your batteries, your devices draw electricity from the grid.

SIMPLE INSTRUCTIONS FOR INSTALLING SOLAR ENERGY IN THE VAN

Each system is composed of some basic components. These components are measured in amperes, watts, or volts. It is easier to think of electric current as water flowing through a tube. Amplifiers measure the volume of water flowing through the tube, volts are like water pressure and watts show how much water the tube could deliver. Use this simple calculation for every part of your solar structure: Ampere x Watt = Volt.

TYPES OF ELECTRIC CURRENT

There are two types of electric current in vans: AC and DC:

- Direct current: The current that is conducted directly by the solar modules. It supplies power to 12V devices and inserts them into 12V batteries. Direct current flows in one direction and requires a positive and negative connection.

- AC power: Energy used in homes to power 120V devices and electrical outlets. AC power is required to charge phones and laptops or to power hairdryers and microwaves. AC pushes the electrons back and forth

You can find many components online, such as 12V. Our domestic electric cooler, MaxxAir fan, potentiometers, and light switches all work with 12V. However, we also have two regular three-pin outlets and a 4-port USB socket that require AC power. An inverter is needed to convert solar energy into alternating current for your devices.

SOLAR COMPONENTS FOR THE INSTALLATION OF SOLAR FIELDS VAN

Every system is different. What you want to feed determines how small or large your system should be. 'Renogy' has an online solar calculator with which it is possible to determine what size of the system is needed and what components are needed to power this system.

When we passed, we knew we wanted to live completely off the net for days and didn't have to worry about being connected to the earth. We knew that we would probably run an InstantPot (900 W) and an induction hob (max. 1800W). I wanted the option to run a mixer that usually rises to 1800W. Most of our devices use very little electricity to operate. In essence, our domestic cooler uses compression technology, for example, and takes less than an amp/hour to keep La Croix cold (including vegetables and other things).

Device performance is essential when looking at an inverter, and ampere-hours are essential when looking at batteries. Batteries as in different capacities of the hour ampere. We have two 200 A batteries now connected to a 2000W inverter/charger, three 100 W solar modules, and a 40 A charge controller.

When installing solar modules for vans in a do-it-yourself solar conversion, there are four main components: battery, charge controller, inverter, and solar modules.

SOLAR FIELDS OF VAN

Your system will be unusable without solar panels! The van solar panels were delivered in different powers, but most van lifts use multiple 100W modules for their systems. The modules themselves are easy to install. In excellent condition, a 100W solar module can produce 30 Ah / day. With 3-100W modules, you should be able to charge a 200 Ah full capacity battery in five sunny hours.

To increase the total energy consumption, it is necessary to add five or six panels to the roof or an additional self-supporting panel that can be removed on sunny days.

Solar panels are available in two types of crystalline silicon; monocrystalline and polycrystalline. Polycrystalline sheets are preferred today. Although a little less efficient (around 4%), they are cheaper to manufacture, making them a favorite of commercial and private buyers.

CHARGE CONTROLLER

The solar panels are connected directly to the truck's charge controller to ensure that they don't overcharge the batteries. The charge controller is an essential part of the solar puzzle. If the solar modules are connected to the charge controller and the charge controller is not connected to the batteries to move electricity, you can detonate the charge controller. There are many warnings on the packaging. When working on the electrical configuration, make sure that the solar panels of your van are mainly disconnected and that the entire system is turned off.

Charge regulators are selected based on their current strength. A solar panel generates around 6 A of electricity per hour. A charge controller with 18 A + would, therefore, be required for three solar modules. A charge controller with 24 A + would be required for four Van solar modules, and so on. We decided on a 40a charge controller in case we want to increase our solar capacity at some point. We could easily add up to three additional panels to our system while maintaining the capacity of our 40 A charge controller.

INVERTER

If you are using a fully 12V system, no inverter is needed. However, if you add appliances or electronic components, you need an inverter in the mix. The inverters take your direct current and exchange it for alternating current. The inverters are measured in watts and must be sized according to the highest peak.

For example, if you want to operate a Vitamix mixer with a power of 1800W, you need an inverter with at least this capacity. It is also essential to consider which appliances you are using simultaneously. If you need to use InstantPot (900W) while operating the induction cooker (1800W), you need an inverter with 3000W +.

Most inverters also have one to four sockets for connecting sockets or appliances. It is also possible to route the 120V outlets through a busbar and return to the inverter cables if preferred.

BATTERY

Three types of batteries are best for van conversions: General assembly, gel, and lithium iron phosphate:

1. AGM battery: AGM stands for 'absorbent glass mat,' which means that floating electrolytes are gathered in glass mats.

2. Gel battery: Gel batteries combine sulfuric acid and silica to form a gelled electrolyte. As a rule, gel batteries do not work well in high heat environments.

3. Lithium iron phosphate battery: Chemically advanced type of battery that lasts longer than gel and AGM batteries.

Each has its advantages and disadvantages. Lithium-ion batteries are the right way for a long-term investment. It is by far the most expensive of the three (often three times the cost), but they have the longest lifespan. AGM and gel batteries can only be used up to 50% of their capacity, which results in a shorter lifespan, around 1000 cycles at 50%. Lithium ions, on the other hand, maintain 1500 cycles at 100%. In essence, gel and AGM batteries have 1/3 of their life cycle as lithium ions.

Batteries are sold in several volts and ampere-hours. In general, 6V batteries are used for small applications and are often found in cars. 12V is the voltage of choice for many 'van-lifers' as it is the standard currency for compression coolers, lamps, and fans.

The ampere-hours indicate how long the battery should work. Since gel batteries only work up to 50% of their full capacity, a 100 Ah battery lasts for around 50 amps. Our 2-200ah batteries are good for about 200 amperes hours. Our compression cooler, which consumes about 1 ampere per hour, can run for two hundred hours if the solar panels do not charge the batteries.

Put it all together

Once you have identified the four main components for the installation of the solar panels of your van, you will have to understand which part is connected to allow a seamless configuration with the van. Once the solar panels are mounted on the roof of your van or camper van, you need to let them pass through the roof. (We used a Blue Sea Systems clamp to prevent water from entering.)

Do not connect the solar modules of the delivery vehicle to the charge controller until the charge controller is connected to the batteries. It could explode!

First, connect the battery to the charge controller. As soon as you do, the charge controller should turn on. Hooray! Otherwise, the batteries may not be charged. Then connect the batteries to the inverter and finally the solar panel cables to the charge controller. At this point, the charge controller should turn on (unless it is dark outside. In this case, the solar modules will not produce any energy).

Make sure the circuit is turned off every time you work on your solar power system. An on / off switch can be purchased in most auto parts stores or on Amazon, that disconnects battery power from the bus bar. And always, always, always disconnect the solar panels from your van first.

If you work with electricity in your van, 1) disconnect the solar panels and 2) turn off the circuit.

Further information on the technical data of the individual products, on the fuses, and on the size of the wire and the thickness can be found in the detailed Renogy instructions.

HOW TO INSTALL SOLAR PANELS FOR STEP-BY-STEP INSTRUCTIONS FOR YOUR SMALL HOME

Solar power plants have improved significantly over the past decade. You have finally reached a point where you can provide all the energy you need for a comfortable off-grid lifestyle.

Today we will cover the basics of sizing your solar system and main components and give you an introduction to putting it all together.

Take the First Steps Towards Solar Self-Sufficiency

If you are planning a small solar system for your home, you should first write down exactly what you need for power. Includes things like:

- Domestic appliances
- laptop

- Cell phones
- lighting
- TV
- Game console
- A / C unit

Once you know everything that uses electricity, you can calculate the amount of solar energy you need.

It is always a good idea to add some margin to this sum in case you need extra energy for anything.

In this way, it is possible to size both the solar modules and the battery bank.

Next, you need to decide whether you want to build your own small home solar system yourself or choose a commercial solar system.

Finished Solar Systems

If you are looking for more plug and play, there are several options available.

By far, the best will come from a company known as 'Goal Zero.'

They manufacture completely independent solar power

generators that combine a charge controller, a battery, and an inverter to create an intuitive device.

In this way, you can connect them directly to Goal Zero portable solar panels or yours.

Yeti 1250

Objective Zero Yeti 1250 is a finished sunlight-based force generator that offers a prepared to use framework from the beginning.

It can be connected directly to all Goal Zero solar modules, but it can also be connected to standard modules via an adapter.

It can take up to 240W of power from solar panels.

The basic system contains a 100 Ah AGM battery.

It is designed for the simultaneous power supply of up to 10 devices and can supply electricity to a standard refrigerator for 20 hours at no additional cost.

This gives it a significant ability in itself, but what makes it truly special is its chaining function.

It is also possible to connect the Yeti 1250 to the other 12V batteries to be charged.

This means that you can have up to several hundred Ah of power simultaneously.

Yeti 1250 is an incredible framework on the off chance that you have restricted force prerequisites and would prefer not to stress over cabling your little house.

It gives you a lot of flexibility and, at the same time, offers you excellent benefits.

The Basics of a DIY Solar Power System

Off-grid solar power systems initially seem intimidating but are fairly simple after dismantling.

Solar panels convert the sun's photons into electricity.

This is connected to a battery bank via a charge controller, which then slides to an inverter and can be used in the home.

There is certainly more to the details, but the bones of the system are easy enough to understand.

A big disclaimer before going any further:

Working with electricity is always associated with risks. If you don't feel 100% comfortable in the wiring of your small house, hiring an electrician is a great idea.

They have special training and know how to properly connect to your solar power system and the rest of the electrical equipment in your small home.

Even if you choose to build the system, you should probably have it checked by an authorized electrician before turning everything on.

The electrician may be able to identify something that you have missed, and that might have had serious consequences if it hadn't been caught in time.

Solar Panels

Solar panels are the biggest and generally most arduous piece of building. You have to situate them for most extreme sun presentation and mount them accurately.

When you install them on a Tiny House On Wheels (or THOW), it is essential to make sure that your accessories can withstand the wind while driving.

When comparing solar panels, there are two necessary numbers to look at:

- Watts
- Solar cell efficiency

A watt is a unit for measuring electricity.

The output of a solar module indicates how much electricity a given module generates in one hour in perfect sunshine conditions. Most modern home panels produce around 100-200 watts per panel.

The efficiency of the cells determines the effectiveness with which a solar panel captures the sun's energy. A panel with 20% efficiency would produce 100 watts of power per square meter. Higher cellular efficiency solar modules produce more electricity due to their size but are generally more expensive.

You also need to decide how to connect multiple solar modules in your solar system. You can connect them in parallel or series.

Parallel Wiring

To connect solar modules in parallel, all positive and negative connections must be connected. This has the effect of further increasing the amplifiers generated by the panels.

The advantage of parallel wiring is its redundancy. If one panel or connector fails, the whole system stops working, and the other panels continue to flow without interruption.

If you are using a PWM charge controller, you will have to wire the panels in this way.

Series Wiring

With a series connection, it is necessary to connect the positive pole of one solar module in the series with the negative pole of another solar module until the entire solar bank has been connected to the charge controller.

This has the effect of further increasing the voltage generated by the panels while the amplifiers remain the same.

The right way of thinking about serial wiring is to use old Christmas lights. The current flows from one panel to another through the system as from incandescent to incandescent.

This has the same disadvantage as the old Christmas lights; if one panel goes out, the whole system no longer works.

The advantage of series control panels is that there are fewer line losses.

This is not a significant consideration for most small residences, but if you are building an off-grid farm with your solar panels, a good distance from your home, you should take that into account.

If you want to use an MPPT charge controller, you must connect the panels in a series.

Charge Controller

A charge controller takes the electricity generated by the solar modules and regulates it to charge the batteries. Batteries are quite sensitive, so it is essential to have a high-quality charge controller in the system.

The charge regulators control the voltage and speed of the battery charge and they also prevent batteries from overcharging and being damaged.

There are two main categories of solar charge controllers on the market today: Pulse Width Modulation (PWM) and Tracking of the maximum PowerPoint.

Each of them works on different principles to charge batteries, but both work in a very similar way.

PWM charge controllers were once a significant step forward, but are now overshadowed by MPPT charge controllers.

PWMs are suitable for small solar power systems, but they are not the optimal choice.

MPPT charge controllers, on the other hand, can generate up to 30% more electricity from solar modules than PWM charge controllers.

One of the biggest disadvantages of DC power is the loss of power due to low voltage transmission.

While PWM charge controllers can only process voltages up to 18V, MPPT charge controllers can process significantly higher voltages and convert them to amps.

This has the effect of limiting the line loss and capturing the additional power that would have been lost.

In summary, MPPT charge controllers are the best option concerning the quality of charge controllers.

Build your Battery Bank

Solar panels are only part of a small solar-powered house. To use the power generated by the panels, you need to store and adjust them.

Sizing of the Battery Bank

When sizing the battery bank, it is essential to take into account the amount of electricity needed between charging cycles. Most batteries have a shorter life span if you increase the depth of discharge (DOD).

This is based on a percentage of their total battery capacity, measured in ampere-hours (Ah). As you increase the depth of discharge, you decrease the number of charge/discharge cycles that the battery has before they fail.

A good rule of thumb is to double your electricity needs. This will give you all the power you regularly need, preserving the battery life.

Types of Batteries

There are tons of different types of batteries, but only lithium and lead batteries work for most domestic and off-grid needs. They maintain the functionality for a good price, which makes them perfect for our purposes.

Lead-acid (Annual General Meeting)

Lead-acid batteries have been in use for decades. The car battery is a lead-acid battery, as are most of the large batteries you have encountered in your life. They work with lead plates suspended in a sulfuric acid solution.

In the past, lead-acid batteries required considerable maintenance and care, but advances in absorbent glass (AGM) batteries have made their job much easier.

These are fixed batteries that require practically no support contrasted with the corrosive lead batteries of the past.

However, they must be ventilated for safety reasons, but it is not necessary to add distilled water.

Lithium Iron Phosphate (LFP)

LFP batteries have become accessible for sun-oriented use at home. They have various advantages over lead corrosive and different kinds of batteries, yet they can be very costly.

Perhaps the best thing about LFP batteries is the permissible release profundity. Up to 80%, the lithium-ion battery can be used regularly.

This gives you significantly higher useful output than a similarly sized AGM battery. They are also lighter and allow many more cycles than AGM batteries.

Many offer a warranty of up to 10 years, compared to a warranty of up to 3 years for AGM batteries. They cost more in advance but are much easier to use and offer higher energy density. If you can afford the initial cost, LFP batteries are always the way to go.

Inverter

An inverter is used to increase the DC power of the batteries to the standard AC power used by modern devices. They are available in two different versions:

- Pure sine wave
- Wave modified or almost sinusoidal

Adjusted sine wave inverters are less expensive and less powerful. They are useful for things like large gadgets, however, they aren't useful for touchy hardware. If you need to charge your cell phone, PC, or TV, you need an unadulterated sine wave inverter.

Unadulterated sine wave inverters are more costly than adjusted sine wave inverters, however considerably more powerful. They allow you to provide almost the same quality of energy as DC batteries that you would get from a normal household outlet.

Switches, Fuses, and Disconnectors

This is very subjective for your design and how much additional protection you want to install. However, there are some components that you should always use to protect your investment. These include:

- Online backup
- Low voltage disconnection (LVD)

- Disconnect switch

Inline fuses must be installed between the solar modules and the charge controller, the charge controller, and the batteries, as well as the batteries and the inverter.

These can be used to protect various system components from a short circuit or overload that triggers the part of the system. The fuse installed in it depends on the size of the solar system and the power of each line. The circuit breakers must be in the same positions so that the system can be completely shut down for maintenance.

Make sure you follow the correct shutdown order:

First, disconnect the solar modules from the charge controller, then the batteries from the inverter, and then the charge controller from the batteries.

A low voltage shutdown occurs between the inverter and the batteries. It prevents the inverter from overcharging the batteries if you drive without the sun for too long.

All these systems must be connected to the positive cable.

Put it all Together

When you connect the different parts of your solar power

system, there is a very specific order in which you should do it.

You also need to make sure that all circuit breakers are set to 'off' when connecting the system.

First of all; never connect the solar modules to the charge controller before having connected the charge controller to the battery bank.

Solar panels should be the last thing that fits into an otherwise complete system. If you connect the panels to the charge controller, you could burn out the system or even cause an explosion in extreme cases.

As soon as you have connected the batteries to the charge controller, it should come on and go through some settings.

Follow the manufacturer's installation instructions and connect the inverter to the batteries.

Only then should you connect the solar modules to the charge controller.

Sustainable Solar Life

With a small home solar power system properly installed, you can go anywhere without being dependent on the power grid. This gives you the flexibility to live in the middle of a city or the middle of nowhere.

HOME SOLAR SYSTEM - DIY SOLAR PANELS

The complete solar energy system for renewable energies consists of do-it-yourself solar collectors and a homemade wind generator.

Alternative energy resources to generate free solar energy to reduce electricity costs.

Many people today save a lot of money on their electricity bills by generating their free solar energy. A lot of information is available online to help you create your solar power system for your home.

In Google's search for homemade wind generators, small solar panels, or an alternative energy source that describes how to generate your electricity, you will find dozens of websites offering a downloadable instruction pack.

Make sure they offer a 60-day money-back guarantee. Fifty dollars would be the highest price to get a good tutorial package with online video tutorials and PDF guides.

By connecting your solar power system to the electricity grid, you can qualify for grid measurement in many states. The grid measurement measures the excess electricity that is generated with the combination of solar panels and a homemade wind generator and feeds it into the local electricity grid.

Solar Power

Solar energy is a free energy source that is renewable and accessible to all homeowners. All homeowners can use solar energy. Even if you can't afford to install commercial solar panels to power your whole house, you can use small solar panels and reduce energy consumption, even if only in a small space.

The money you save on electricity bills can be invested in building additional solar panels and further reducing electricity, as well as saving more money until you reach a stage where excess electricity is generated compared to what you consume. You will be credited with excess electricity, and the energy supply company will start paying you the money.

Fossil fuels will not be able to sustain our energy consumption if the world looks to the future at the speed with which we use this source of energy. This alone is a good reason to build a solar system for home use. Even the smallest solar generator used to power your laboratory or external safety lights is a start to saving our planet and the environment.

By generating solar energy, we produce less toxic fumes and use fewer chemicals that are by-products of the energy sources we use today. Solar power generators and small solar modules require very little maintenance to keep them fully functional for many years.

A do-it-yourself solar collector or grid-connected home solar system doesn't have to be an expensive exercise, as you can start building your home solar system for less than $180. A complete, interconnected solar system for private homes costs more and depends on the number of solar modules installed.

With so many packages available on the internet for instructions on how to make a small solar power generator, you can be sure of getting the right information. Still, you have to be very careful with what you choose, as there are many energy packages out there that are garbage. Complete.

You don't have to be an electrician and use expensive machinery or tools to build your home solar system. Most likely, the material is easily accessible from your landfill and hardware store.

Most of the available packages are easy to read, easy to follow, and have fully illustrated instructions on how to set up and install your home solar power system. Most internet packages explain the secrets to finding cheap solar cells.

Wind Energy.

Building a homemade wind generator for home use quickly becomes an interesting option for many environmentally conscious homeowners to reduce both environmental damage and the consumption of fossil fuels as an alternative energy source.

Homemade wind generators are becoming more common as homeowners install them on large and small solar power systems. A wind generator converts the wind into a source of domestic energy, which is free, unlimited, renewable, and green!

The wind generator converts physical movement and natural wind to convert the blades that cause rotation into electricity using powerful permanent magnets. Freely generated domestic electricity can then be connected to devices for immediate use, stored in batteries, or fed back into the mains.

Two things that are very important before going too far into a homemade wind power generation system for your home are:

1. It is necessary to live in a windy area for a wind generator to work properly at home and to integrate the energy into the needs of the home. Domestic wind turbines need a lot of wind to function properly. Otherwise, you have to keep up with solar energy.

2. The domestic wind generator can only be an addition to the domestic solar energy system.

You're probably trying to decide if you can learn how to build a DIY wind turbine, so you'd like to know ...

- How much does it cost to build a DIY wind turbine?

- Can I get reliable and easy-to-follow plans for the wind generator?

- How long will it take to build a self-made wind generator?

- Is a self-made wind generator comparable to commercial models?

- Is it easy to find parts for building a homemade wind turbine?

Yup! - You can build a DIY wind generator for less than $180 in about two weekends by following the directions in the packages.

Renewable energy [wind energy] works on the same basis as

hydroelectric energy, the difference being that it is driven by the wind and not by water.

Two DIY projects are relatively easy to install to deal with the energy crisis. Regardless of whether you use solar energy for the sun or wind power, you conserve our natural resources, the planet, and the environment.

Many illustrated guides are available online to teach you to step by step on how to make small DIY solar panels and wind turbines.

How to create a complete solar system for your home with all the solar collectors you want to install, depending on your financial situation, to be partially or completely disconnected from the grid.

If you want to do it yourself, these guides will make your home less dependent on fossil fuels and more on renewable energy, while reducing electricity bills.

Assuming an average home pays over $200 a month, that's an annual cost of around $2400 for your electricity. If you install a solar system for your home, you will cut down on electricity and save money.

Use natural gas for heating or cooking. This could result in energy savings of several hundred dollars per year. Outdoor cooking is a lifestyle in some countries, and you can adapt to it to reduce the electricity needs of your solar system at home. This means fewer solar panels, which reduces the production and installation costs of the solar system.

With a do-it-yourself approach to solar systems at home, you can help support the environment on Earth and remove part of your carbon footprint.

TEN REASONS TO INSTALL A SOLAR-POWERED PUMPING SYSTEM

Have you at any point contemplated introducing a solar well pump? Do you have a rural or remote location where you have to pump water from the surface or several hundred meters below? A solar well pump is a perfect solution.

Technological developments in both pumps and solar energy have made this possible. Both solar panels and solar pumps have made progress, enabling them to meet a wide range of water pump requirements. Many of these systems did not exist until a few years ago.

The ten most important reasons for installing a solar well pump are listed below. If you've never thought about installing a solar system, look in the list and see if you can imagine using it. It is surprising when ideas for cheap remote water pumping are made available, ideas, and uses come to mind! Don't hesitate and look for a system today.

1. Pump water around the world. No external power supply required.

The problem with pumping water in rural areas is the need to direct electricity to the site. Wind power and windmills have been used in these remote locations for many years. Windmills are expensive and difficult to maintain. There are better options today.

The turning point in pumping solar water is that no external energy source is needed. The solar collectors provide all the electricity needed to pump water from several hundred meters below.

2. Solar pumps are more efficient and powerful than ever.

Today's solar pumps are not like the pumps of the past. These are powerful, efficient, and commercial products. First-class solar fountain pumps are made of stainless steel and have brushless DC motors.

Stainless steel is used for the siphon lodging and the siphon component to guarantee legitimate cleanliness and ensure long life. Stainless steel resists corrosion, even if it has been suspended in water for years. The mechanisms of the anti-rust pump minimize wear from sand and other particles and lift water from the deep subsoil.

Brushless DC engines are among the most productive available. Since they are brushless, they must never be removed from the well to change the brushes. These engines are classified for tens of thousands of hours of maintenance-free operation.

3. The systems are cheap and readily available.

Past solar power plants had high prices in the tens of thousands of dollars. Advances in innovation have made systems possible and effectively accessible. One of the most important advances that made this possible was the solar cell and solar panel. The production of solar cells has progressed so much that it is now very convenient. A system that used to cost tens of thousands of dollars is now in the low thousands. A 10-fold reduction!

A simple but complete pumping system costs around $2,000. This basic system pumps water from a depth of several hundred feet to a flow rate of a few liters per minute. This basic system replaces most windmills and regulates both depth and flow. They will fill a large pond with water without operating costs or pump enough water for a few hundred cattle.

Higher performance systems raise the cost to around $3-4,000 because more depth and reach are needed. These systems are suitable for supplying water to entire families or hundreds of farm animals. The standard home fountain pump can be replaced. There is a slight increase in the cost of these systems due to higher performance, but remember, these are complete systems, including solar panels (high cost).

Specialized systems can cost tens of thousands of dollars. These are high-performance systems that pump many liters of water per second, which is enough for an entire farm or even a remote village. They are exaggerated for most livestock and household needs. They are best suited for large farms and crop irrigation.

4. No running costs.

After the initial cost of the system, which is often compared to other downhole options, there are no ongoing operating costs. Normal wells burn money every time they light up to pump water. This is not the case with good solar pumps that absorb their electricity from the sun.

Every day the sun shines, you can earn money with the sun for free. What better deal is there? The sun shines, and the water is pumped out of the deep subsoil, protecting you from expensive electrical costs.

5. Much cheaper than installing mains power in remote

locations.

Shutting down the mains power in a remote location takes time and money. This is one of the reasons why past windmills were so popular because they didn't need electricity to operate. The 'windmills of today' are solar energy systems that can pump water without being connected to the main power supply. You are not, at this point, constrained to what extent a force rope can be covered in the ground or how strong the wind blows.

6. Complete systems can be delivered directly to your home.

Complete systems can be delivered to your home in a few days. Only two boxes are needed, one for the pump/controller and one for the solar panels to get directly to your home or office. Many standard systems can be shipped using normal shipping methods (USPS, UPS, FedEx) without the need to ship goods. This allows shipping systems in all locations and all residences.

7. The systems are modular and can be updated over time.

The components that make up the pumping systems of solar energy wells are very modular. They can be exchanged and updated as needed. If more cloud energy is needed, additional solar panels can be added. If more water is needed per day, additional batteries and panels can be added to allow water to be pumped overnight. If a larger volume is needed, the pump can easily be replaced with a high-performance model that offers higher flow.

8. Solar good pumps are easy to maintain.

There are very few mechanical components. Solar panels are very reliable and require no maintenance other than a quick wash every two years. Brushless motors have no brushes (hence the name) and therefore require no maintenance. The pump mechanisms are very reliable and simple to supplant in the field. A framework ought to have the option to siphon a huge number of gallons without maintenance.

9. Easy to maintain.

Two elements in solar pumping systems occasionally require maintenance. Both are very cheap. Occasionally, solar modules need to be washed up to two every years to ensure maximum performance. Less often, about every five years, it makes sense to replace the pump mechanisms, which can deteriorate over time and affect performance. These mechanisms cost around $20 and can be replaced on the spot.

10. The systems can be installed alone in one weekend.

Solar energy well pumping systems are very simple and can be installed in just one weekend. All solar panel connections are made with waterproof connections without the need for welding. The solar pump, solar modules, and sensors are all connected to the controller via screw terminals, no welding is required here. The pump should be easily connectable to existing cables and supply pipes. Lower it into the well, and if you've completed the installation before the night, the water would flow!

There you have the ten most important reasons to install a solar energy well pumping system. Yes, there are still hundreds of them. Don't hesitate and install your system today!

www.ingramcontent.com/pod-product-compliance
Lightning Source LLC
Chambersburg PA
CBHW070240220526
45465CB00004B/1459